100 Geometric Games

PIERRE BERLOQUIN

100 GEOMETRIC GAMES

Foreword by Martin Gardner

Drawings by Denis Dugas

CHARLES SCRIBNER'S SONS · NEW YORK

Copyright © 1976 Pierre Berloquin
© Librairie Générale Française, 1973

Library of Congress Cataloging in Publication Data
Berloquin, Pierre.
  100 geometric games.
  Translation of 100 jeux geometriques.
  1. Mathematical recreations.   2. Geometry—Problems,
exercises, etc.   I. Title.
QA95.B4513   793.7′4   75-40458
ISBN 0-684-14611-8

1 3 5 7 9 11 13 15 17 19 V/C 20 18 16 14 12 10 8 6 4 2

Printed in the United States of America

Contents

FOREWORD

Pierre Berloquin, who put together this stimulating and delightful collection of mind benders, is a clever young Frenchman who was born in 1939 in Tours and graduated in 1962 at the Ecole Nationale Supérieure des Mines in Paris. His training as an operations research engineer gave him an excellent background in mathematics and logical thinking.

But Berloquin was more interested in writing than in working on operations research problems. After two years with a Paris advertising agency, he decided to try his luck at free-lance writing and this is how he has earned his living since. In 1964 he began his popular column on "Games and Paradoxes" in the magazine *Science et Vie* (Science and Life). Another column, "From a Logical Point of View," appears twice monthly in *The World of Science,* a supplement of the Paris newspaper *Le Monde.* Occasionally he contributes to other French magazines. One of his favorite avocations is leading groups of *"créativité,"* a French cocktail of brainstorming, synectics and encounter therapy, for the discovery of new ideas and the solution of problems—a logical extension of his interest in puzzles.

Berloquin's published books are *Le Livre des jeux* (card and board games), *Le jeu de Tarot* (Tarot card game), *Testez votre intelligence* (intelligence tests), *100 grandes réussites* (solitaire games), *Un souvenir d'enfance d'Evariste Galois* (Memoir of the Childhood of Evariste Galois), and *100 jeux de cartes clas-sigues* (card games); he is co-author of *Voulez-vous jouer avec nous* (Come Play with Us) and *Le livre des divertissements* (party games).

This volume is Berloquin's own translation into English of one of his four paperback collections of brainteasers which have been enormously popular in France and Italy since they were published in Paris in 1973. This one is concerned only with geometrical puzzles. The other three contain numerical, logical, and alphabetical problems. Denis Dugas, the graphic artist who illustrated all four books, is one of the author's old friends.

The puzzles in this collection have been carefully selected or designed (many are original with the author or artist) so that none will be too difficult for the average reader who is not a mathematician to solve, and at the same time not be *too* easy. They are all crisply, clearly given, accurately answered at the back of the book, and great fun to work on whether you crack them or not.

At present, Berloquin is living in Neuilly, a Paris suburb, with his wife, Annie, and their two children.

MARTIN GARDNER

PROBLEMS

Game 1

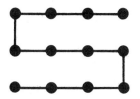

Twelve points are connected above by five straight lines, without raising the pencil.

You can do better: connect the same twelve points:

- without raising your pencil
- in five straight lines
- ending on the first point, thereby making a closed circuit
- without going through any point twice (but the lines can cross each other).

How?

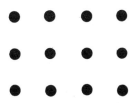

Game 2

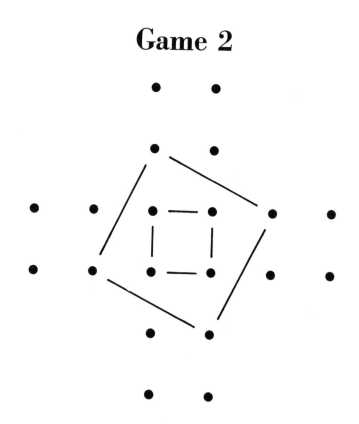

Only two squares are shown out of many squares whose vertexes lie on four of the twenty points in the figure.

How many points do you have to erase so that *no square* can be formed on any four of the remaining points?

Game 3

Is the rope a simple loop?

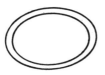

or is it knotted once?

or is it knotted several times?

Game 4

One of two identical coins remains motionless while the other coin rotates around it, touching it without slipping.

When the second coin has completed a turn around the first coin, how many turns has it made around its own axis?

(Solve the problem without using actual coins.)

Game 5

Go through the maze.

Game 6

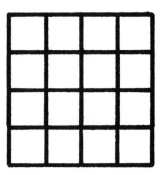

Can you place four chess queens on the board so that none of them threatens another?

(A queen can move any number of squares horizontally, vertically, or diagonally.)

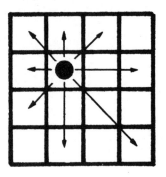

Game 7

How can you go through the garden:

- going along each walk once and only once
- without crossing your path
- finishing at your starting point?

Game 8

The sixteen matches form five squares. Can you change the position of three matches so that only four squares are formed by the sixteen matches?

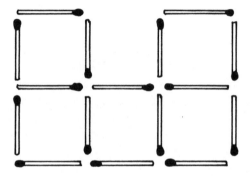

Game 9

How many triangles are there in the diagram? Can you count them methodically enough not to miss any?

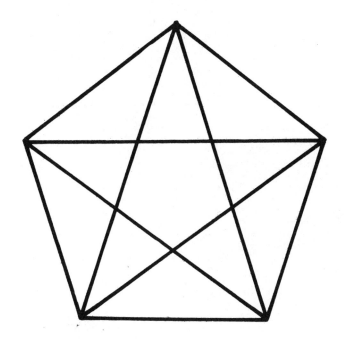

Game 10

Among the six drawings
five are identical, but
rotated by multiples
of sixty degrees.

The sixth drawing
is different.
Which one is it?
Why?

Game 11

Traverse the maze before you proceed on your journey.

Game 12

How many triangles are there in the diagram?

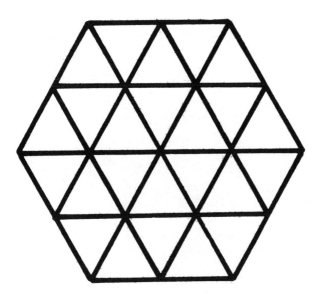

Game 13

A snail has undertaken to climb a pile of ten bricks. It can climb four bricks in an hour. But then, since the effort has been extremely tiring, it must sleep an hour, during which it slips down three bricks.

How long will the snail take to reach the top of the pile?

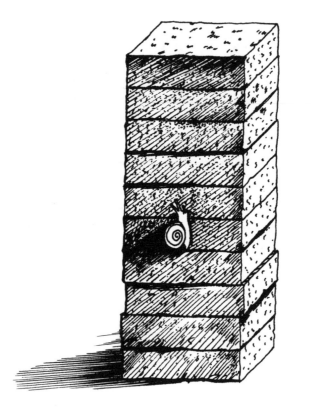

Game 14

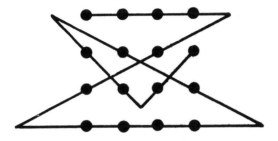

Sixteen points are connected above by six straight lines, without raising the pencil and without going through any point twice.

You can do better: connect the same sixteen points:

- without raising your pencil
- in six straight lines
- without going through the same point twice (but the lines can cross each other)
- ending on the first point, thereby making a closed circuit.

How?

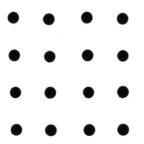

Game 15

The side of the small square is one meter and the side of the large square one and a half meters.

One vertex of the large square is at the center of the small square.

The side of the large square cuts two sides of the small square into one-third parts and two-thirds parts.

What is the area where the squares overlap?

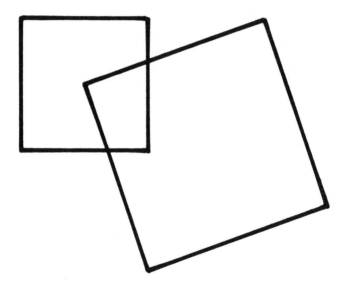

Game 16

How many knots are on this rope?

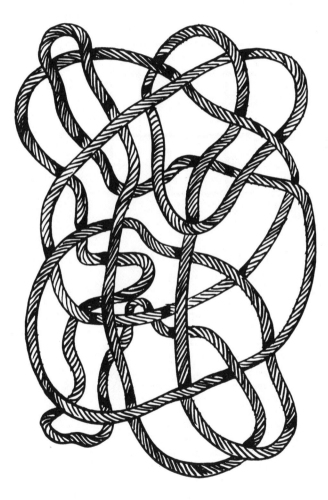

Game 17

Place five chess queens on the board so that none of them threatens another. There are two independent solutions.

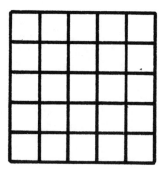

Game 18

Which two matches should you remove so that only two squares are left?

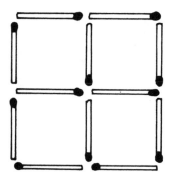

Game 19

Where should you start and where should you finish to go through the garden:

- going along each walk once and only once
- without crossing your path?

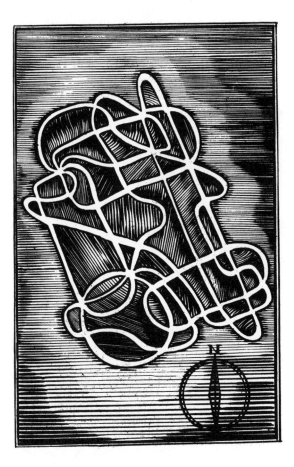

Game 20

Among the six drawings
five are identical, but
rotated by multiples
of sixty degrees.

The sixth drawing
is different.
Which one is it?
Why?

Game 21

Can you reach
the center of
the maze?

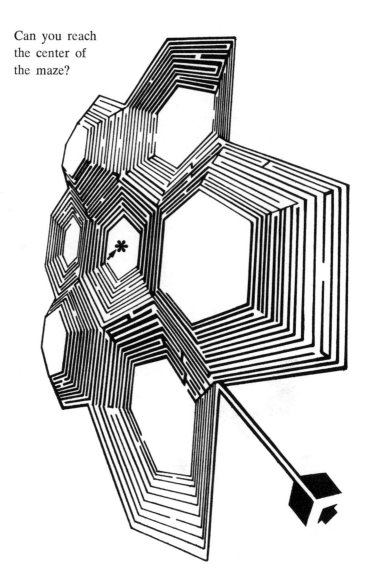

Game 22

How many regular hexagons are there in the diagram?

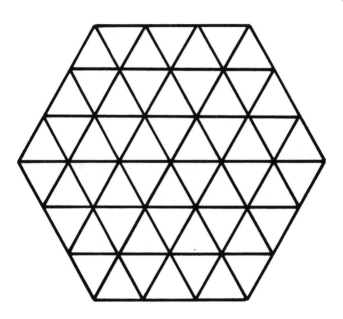

Game 23

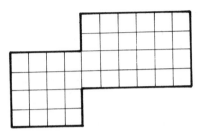

The figure above can be cut to make two identical parts.

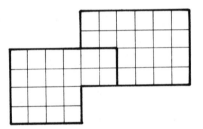

Can you cut the figure below to make two identical parts?

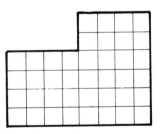

Game 24

An air squadron has about fifty planes. Its flight pattern is an equilateral triangle; every plane except the first is halfway between two planes ahead of it. Several planes are shot down in combat. When the squadron returns, the planes form four equilateral triangles. The lost planes could have formed another equilateral triangle.

If all these triangles are different in size how many planes were there to begin with?

Game 25

From three points you can form three rows of two points each.

Can you arrange ten points to form five rows of four points each?

Game 26

If you enter this garden through its door, how can you go through it:

- going along each walk once and only once
- without crossing your path?

Game 27

There are two loops in the rope. Are they independent?

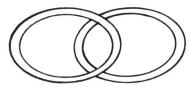

Or are they interlocked?

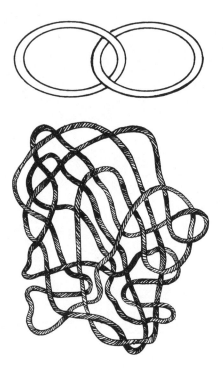

Game 28

Which six matches should you remove, without changing the position of the others, so that only three squares are left?

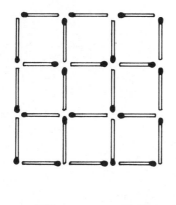

Game 29

There is already one chess queen on the board. Place five more so that none of the six queens threatens another.

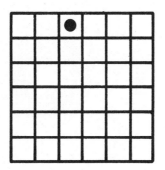

Game 30

Among the six drawings
five are identical but
are rotated by multiples
of sixty degrees.

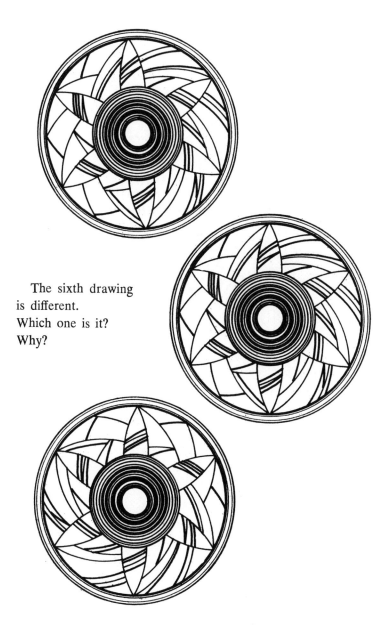

The sixth drawing
is different.
Which one is it?
Why?

Game 31

Enter the maze, and exit as shown.

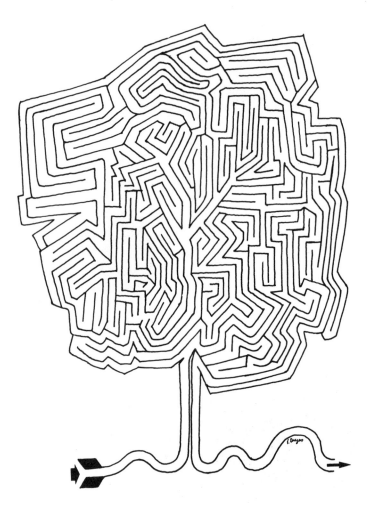

Game 32

How many triangles are there in the diagram?

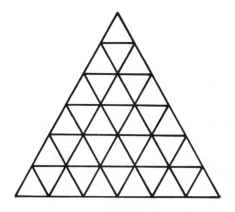

Game 33

Cut the figure to make two identical parts.

Game 34

Place eight chess queens on the white squares of the board so that none of them threatens another.

One queen is already on the board.

The queens cannot be placed on black squares, but they can move through them.

Game 35

Timothy wants to saw a cube of wood into twenty-seven equal cubes.

During the work, if several pieces are already sawed, he can arrange them as he pleases, then saw through all of them with one cut.

Working this way, how many operations are needed?

Game 36

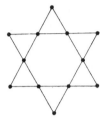

From twelve points you can form six rows of four points each.

Can you rearrange the points to keep six rows of four with only two rows parallel?

Game 37

There are three loops in the rope. How many are independent? How many are interlocked?

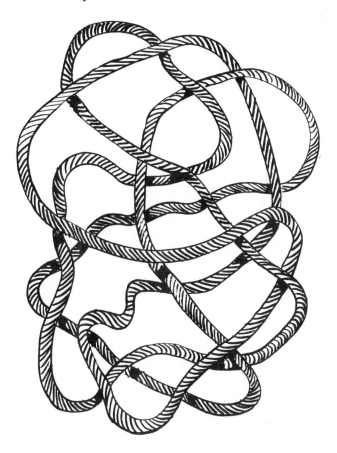

Game 38

Where should you place an odd number of matches inside the square so that four lots of equal area are fenced off?

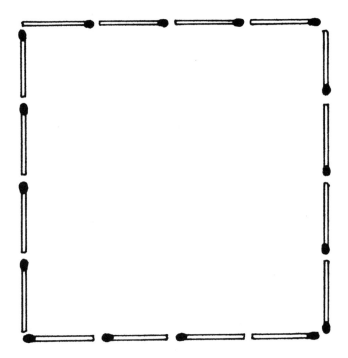

Game 39

Can you go through the forest:

- going along every road once and only once
- without crossing your path
- finishing at your starting point?

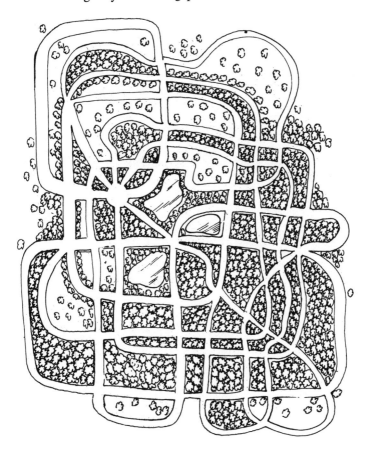

Game 40

Among the six drawings
five are identical but
are rotated by multiples
of sixty degrees.

The sixth drawing
is different.
Which one is it?
Why?

Game 41

How can you go from one eye to the other of this owl-maze?

Game 42

How many rectangles are there in the diagram? (Note: A square is a rectangle.)

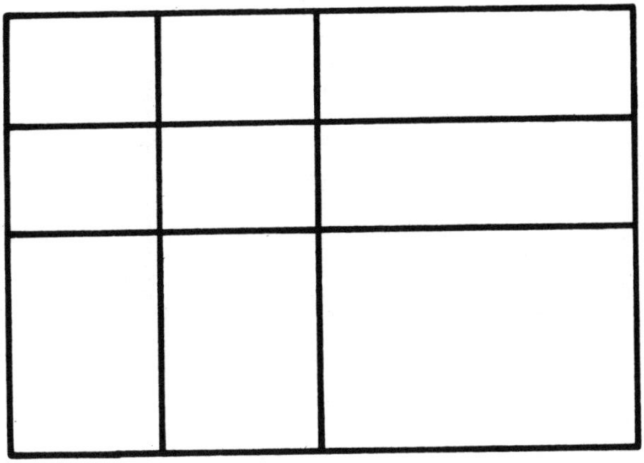

Game 43

Cut the figure to make two identical parts.

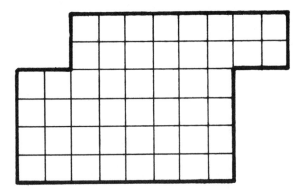

Game 44

Can you arrange thirteen points to form twelve rows of three points each?

Game 45

The bicycle is stationary on the ground; its tires do not slide. A man kneels by the bicycle and pulls the bottom pedal backward (arrow).

Will the bicycle go forward or backward?

Game 46

Ten coins are in a row, five heads on the right and five tails on the left.

In as few moves as possible, we want to alternate heads and tails.

The only move permitted takes two consecutive coins and places them in a two-coin-wide space in the same order they were picked up in. If there is no space between coins, the two coins can be placed at one of the ends of the row.

For example, here are the first and second moves of one attempt.

Game 47

How many loops are there? How many are free? How many are interlocked?

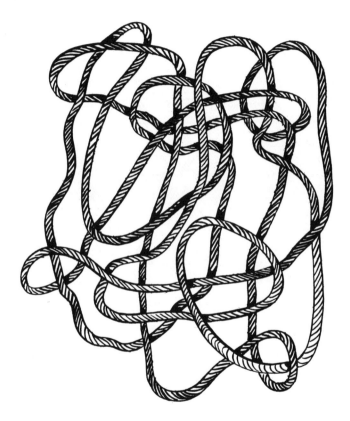

Game 48

Can you change the position of four matches so that exactly three equilateral triangles are formed? (Don't remove any matches.)

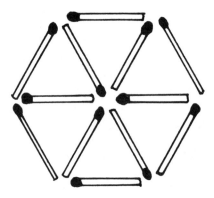

Game 49

You will discover that it is impossible to go through the garden:

- going along each walk once and only once
- without crossing your path.

Only one short walk has to be added to make it possible. Where?

Game 50

Among the six drawings
five are identical but
are rotated by multiples
of sixty degrees.

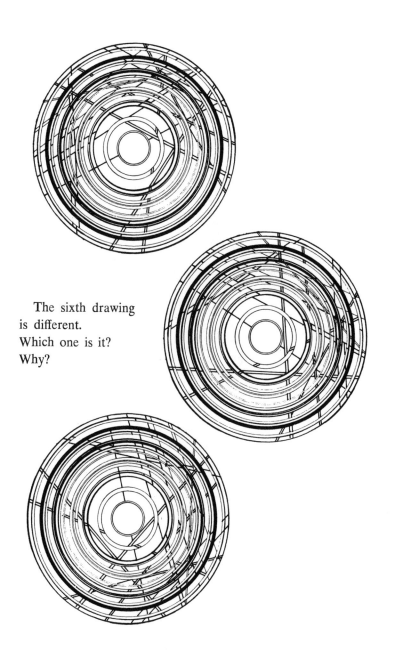

The sixth drawing
is different.
Which one is it?
Why?

Game 51

Here is a maze of a new kind. It has three dimensions and its roads must be followed logically, even when they disappear momentarily from sight under other roads.

Can you get from one side of the maze to the other?

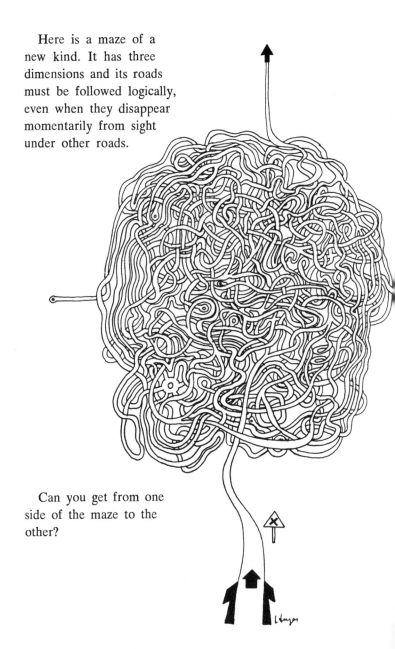

Game 52

How many triangles are there in the diagram?

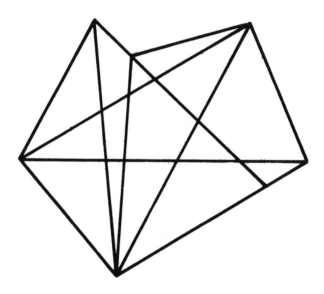

Game 53

How many rectangles are there in the diagram?

Game 54

Cut the figure to make two identical parts.

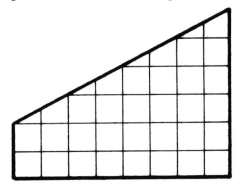

Game 55

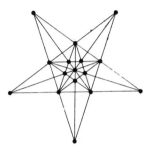

From sixteen points you can form fifteen rows of four points each.

Can you rearrange the points so they form ten rows of four, and no two rows are parallel?

Game 56

Nine glasses are in a row, all right side up. We want them all upside down.

A permissible move reverses any six glasses, putting each one upside down if it is right side up, or right side up if it is upside down.

For example, we start with

A first move reverses the six glasses on the right.

A second move reverses the six glasses on the left.

A third move reverses glasses 2 to 7.

And so on . . .

Can you get all the glasses upside down? How many moves does it take?

Game 57

How many loops are there? How many are free? How many are interlocked?

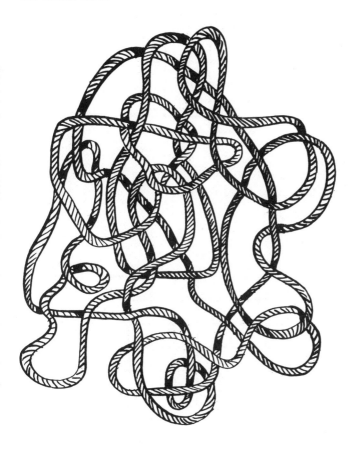

Game 58

A chess knight threatens eight squares at most.

How many knights do you have to place on the 8 x 8 chessboard so that each square is:

- occupied by a knight
- or threatened by at least one knight?

Game 59

What is the smallest number of matches you can remove so that no square of any size is left?

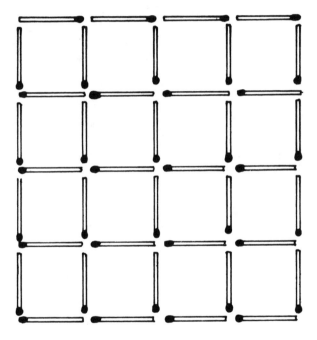

Game 60

Among the six drawings five are identical, but rotated by multiples of sixty degrees.

The sixth drawing
is different.
Which one is it?
Why?

Game 61

Do you know how to cross this maze? It is in three dimensions (roads may disappear momentarily from sight under other roads) and you must obey this rule: on entering a traffic circle, exit by either the first road on your right or the first road on your left.

Game 62

How many quadrilaterals are there in the diagram? A quadrilateral is *any* four-sided figure. (Beware—there are more than ten.)

Game 63

How many quadrilaterals are there in the diagram?

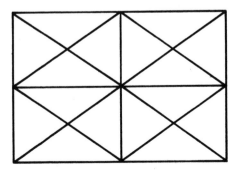

Game 64

How many loops are there? How many are free, interlocked, knotted?

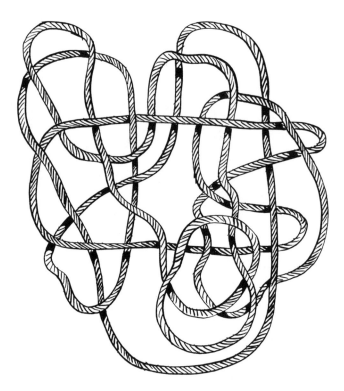

Game 65

Cut the figure to make two identical parts.

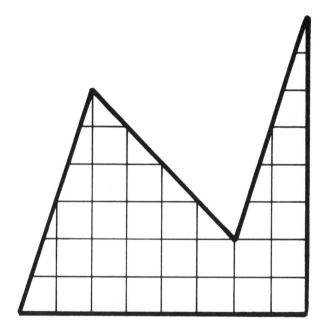

Game 66

A monkey weighing fifty kilograms is climbing a rope. The rope goes over a pulley and is fastened on the other side to a fifty-kilogram weight. The pulley rotates without friction around a fixed axis.

The monkey is doing enough work to climb forty centimeters per second if the rope was fixed.

Does the monkey go up or down? How fast?

Game 67

How many straight lines are needed to separate each star from all the others? Draw them.

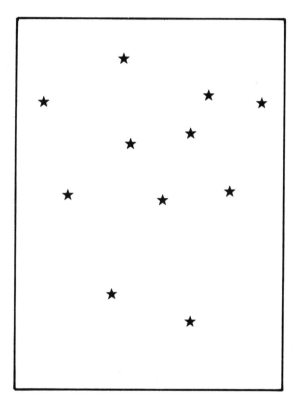

Game 68

Two knights are already on the chessboard. How many knights do you have to add so that each square is occupied or threatened by a knight?

Game 69

Can you change the position of four matches in this spiral so that exactly three squares are formed? (Use all the matches.)

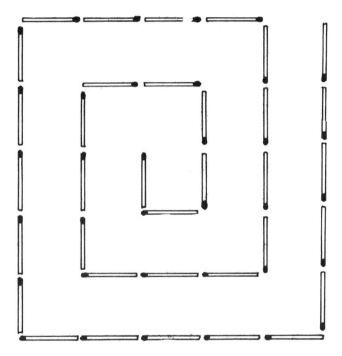

Game 70

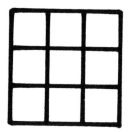

In this diagram, eight equal line segments (four horizontal and four vertical) form 14 squares:

$$1 \times 1 \quad 9 \text{ squares}$$
$$2 \times 2 \quad 4 \text{ squares}$$
$$3 \times 3 \quad 1 \text{ square}$$

Can you rearrange the eight line segments in one diagram so you have:

2 squares

24 isosceles triangles?

The squares are of different sizes. There are 4 big triangles, 8 of intermediate size, and 12 small ones.

Game 71

It is possible to cross this maze made of pipes. How?

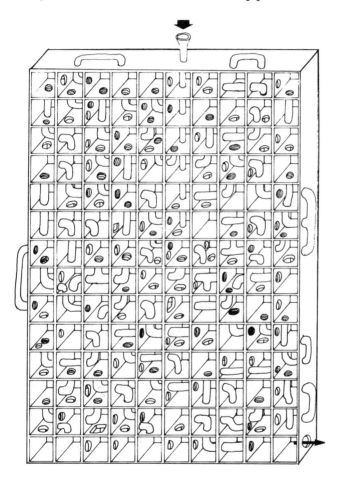

Game 72

How many triangles are there in the diagram?

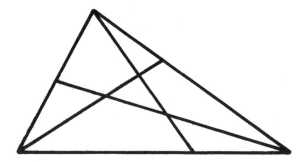

Game 73

Cut the figure to make two identical parts.

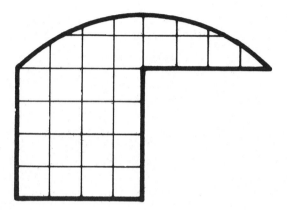

Game 74

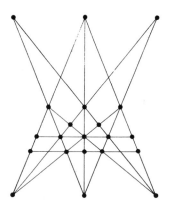

From twenty-one points you can form eleven rows of five points each.

Can you arrange the points to form twelve rows of five?

Game 75

How many straight lines are needed to separate each star from all the others? Draw them.

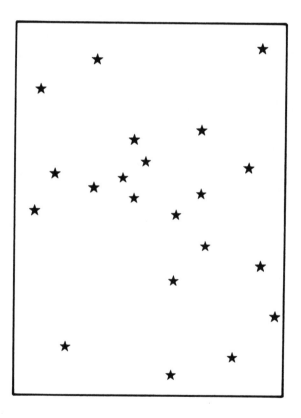

Game 76

How many loops are there? How many are free, interlocked, knotted?

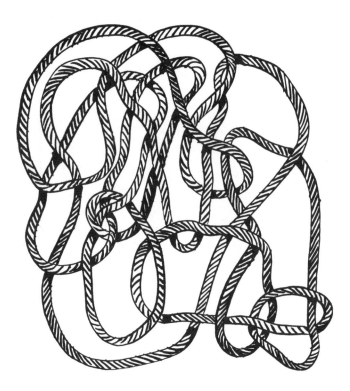

Game 77

How many bishops do you have to place on a chessboard so that each square is:

- occupied by a bishop
- or threatened by at least one bishop?

(A bishop can move any number of squares diagonally.)

Game 78

The two enclosures are made of twenty matches. Using all the matches, can you form two new separated enclosures so that one area is three times the other?

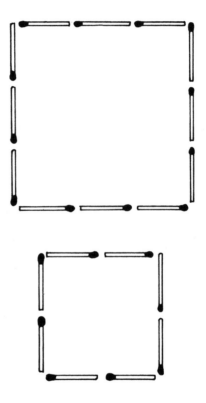

Game 79

A chess knight is in a corner of the board, ready to tour it in a series of moves, occupying each square once and only once and finishing where it started.

Actually, the tour is impossible. Why?

Game 80

Among the six drawings five are identical, but rotated by multiples of sixty degrees.

The sixth drawing
is different.
Which one is it?
Why?

Game 81

Can you go through the maze of this electronic circuit?

Game 82

How many hexagons, regular or not, are there in the diagram? Crossed hexagons (that is, hexagons with sides that continue through an intersection) aren't allowed—which still leaves more than three hundred hexagons . . .

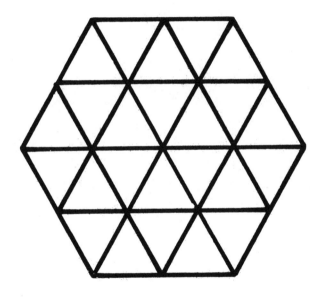

Game 83

How many loops are there? How many are free, interlocked, knotted?

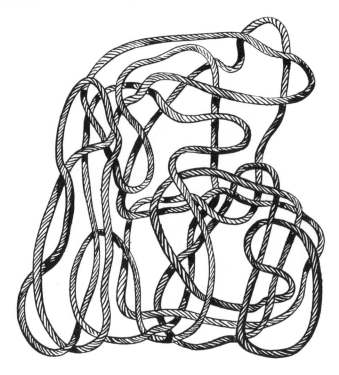

Game 84

Timothy, Urban, and Vincent are running the hundred-meter dash.

Timothy and Urban will reach the tape together if Timothy is given a head start of twenty meters. Urban and Vincent will reach the tape together if Urban is given a head start of twenty-five meters.

Timothy and Vincent want to reach the tape together. Who gets a handicap, and how much? (Assume each man always runs at the same speed.)

Game 85

Can you arrange twenty-two points to form twenty-one rows of four points each?

Game 86

Fill each square with one of the five symbols so that the same symbol does not appear twice:

- in any horizontal row
- in any vertical column
- in either of the two main diagonals.

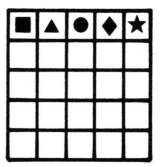

Game 87

How many bishops do you have to place on the chessboard so that each square is threatened by at least one bishop. (If it is occupied, it must be threatened by at least one other bishop.)

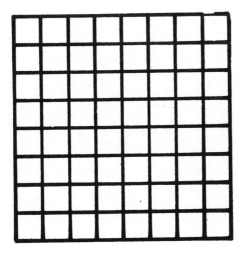

Game 88

Using twelve matches, can you form a quadrilateral with the same area as this rectangle?

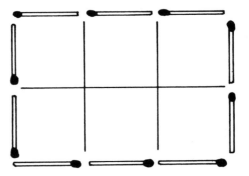

Game 89

How many loops are there? How many are free, interlocked, knotted?

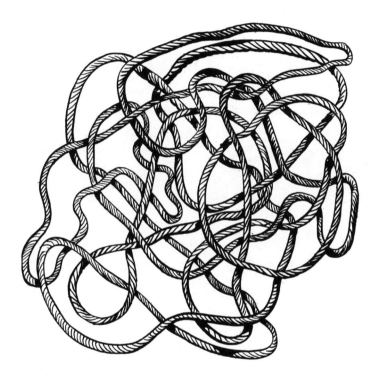

Game 90

Among the six drawings five are identical, but rotated by multiples of sixty degrees.

The sixth drawing
is different.
Which one is it?
Why?

Game 91

Beware! Experimental maze, in three dimensions (paths may disappear momentarily from sight under other paths). You may have to scale vertical walls to do it, but it is possible to reach the uppermost terrace. How?

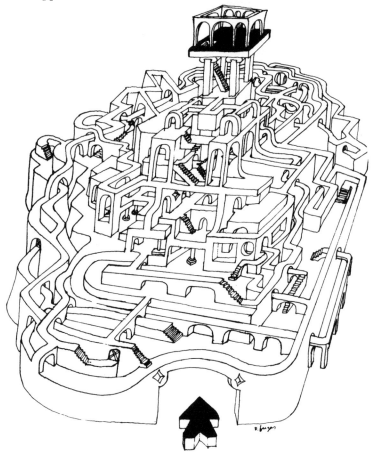

Game 92

How many loops are there? How many are free, interlocked, knotted?

Game 93

Can you cut the vase in three pieces and assemble them to form a square? Note that the vase is entirely composed of curves.

Game 94

The board has forty-nine squares. One square is marked with a star. Can you cover the forty-eight remaining squares with twenty-four two-square dominoes?

Game 95

On the chessboard place fifty-one pieces:

- eight queens
- eight rooks
- fourteen bishops
- twenty-one knights

so that no queen threatens a queen, no rook threatens a rook, no bishop threatens a bishop, and no knight threatens a knight.

Of course, two pieces cannot occupy the same square; but in this game, pieces can move through squares occupied by other kinds of pieces.

(The moves of all the pieces except the rook have already been defined. A rook can move any number of squares horizontally or vertically.)

Game 96

How many quadrilaterals are there in the diagram?

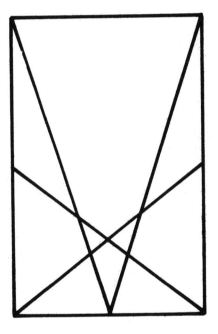

Game 97

How many loops are there? How many are free, interlocked, knotted?

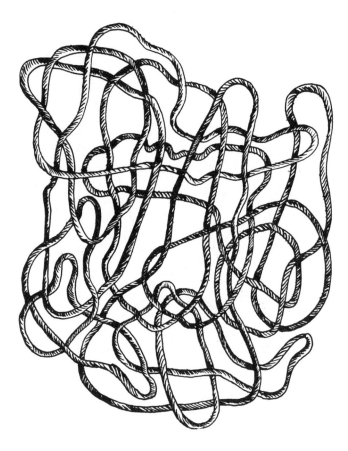

Game 98

A rectangular tiled floor in Timothy's house has 93 square tiles on the short dimension and 231 on the other.

Timothy draws a diagonal from one corner to the opposite corner. How many tiles does it cross?

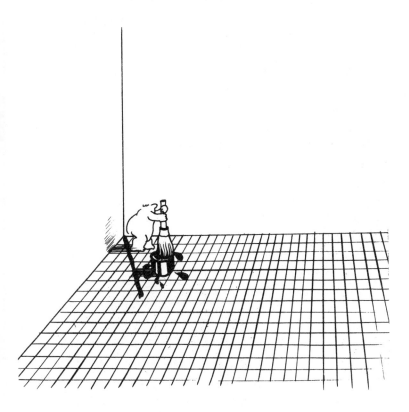

Game 99

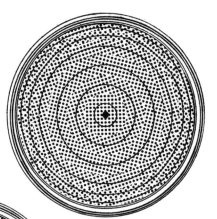

Among the six drawings five are identical, but rotated by multiples of sixty degrees.

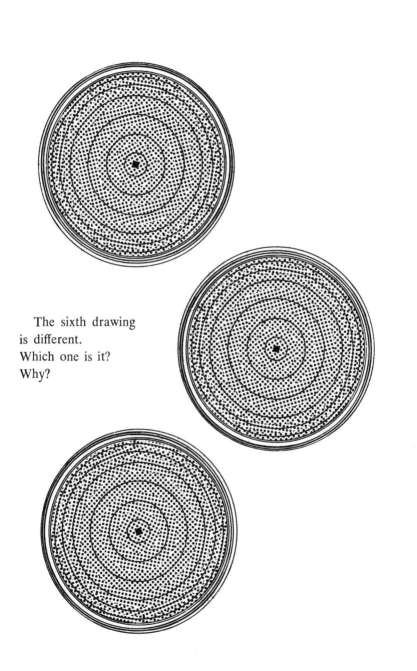

The sixth drawing
is different.
Which one is it?
Why?

Game 100

This maze contains some one-way streets. You only have to cross the bridge between the two halves of the maze once, but

if you find yourself returning on it, don't give up: keep going. (Note: You can go on roads under the bridge.)

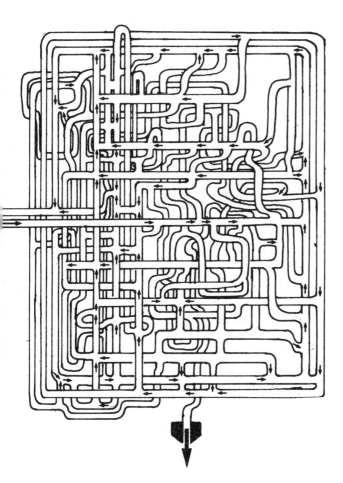

SOLUTIONS

Game 1

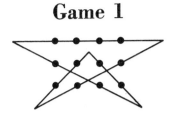

Game 2

You must erase six points.

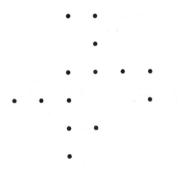

Game 3

The rope is knotted twice.

Game 4

The second coin makes two revolutions:

- one because it rolls on a path as long as its circumference
- one because the path is closed.

Game 5

Game 6

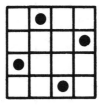

Game 7

There are many solutions. Here is one.

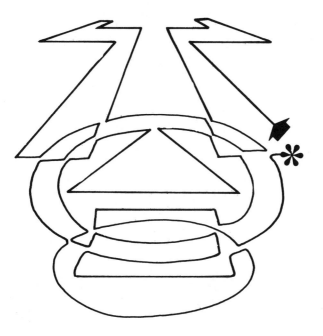

Game 8

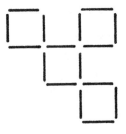

Game 9

The interior lines cut the diagram in eleven elementary areas. Count the triangles according to the number of elementary areas they contain.

1 area	10
2 areas	10
3 areas	10
5 areas	5
	35

There are thirty-five triangles.

Game 10

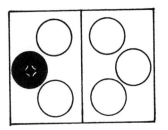

Game 11

Game 12

Classify the triangles by the number of elementary triangles they contain.

1 triangle	24
4 triangles	12
9 triangles	2
	38

Game 13

The snail climbs the pile with an average speed of one brick every two hours. But it does not need twenty hours to climb ten bricks.

At the end of the twelfth hour, the snail wakes up, fresh and rested, at the top of the sixth brick. Then it can spend the thirteenth hour climbing the four last bricks to the top.

Game 14

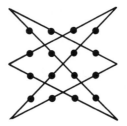

Game 15

Prolonging two sides of the large square cuts the small square in four equal parts. Since the area of the small square is one square meter, the area of overlap is a quarter of a square meter.

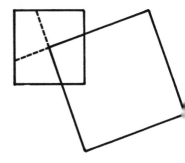

Game 16

There are four independent knots.

Game 17

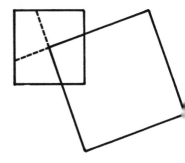

Game 18

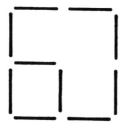

Game 19

It is possible to start or finish on either of two points, which are the points where an odd number of walks meet. Pairs of walks are used to avoid them. Single walks are used to reach them or leave them.

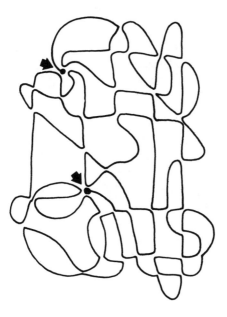

Game 20

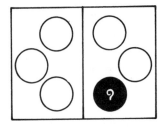

Game 21

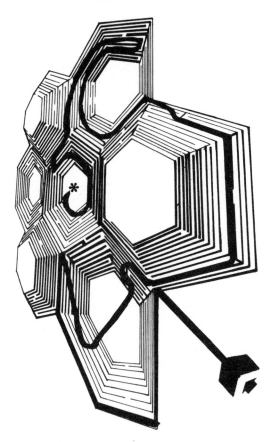

Game 22

Classify the hexagons by the number of elementary triangles they contain.

6 triangles	19
24 triangles	7
54 triangles	1
	27

Game 23

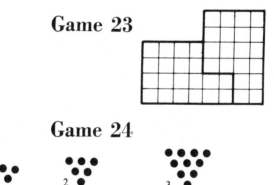

Game 24

The smallest equilateral triangle contains three planes in rows of one and two. The second possible triangle is formed of 1+2+3 = 6 planes. On the same principle, the third, fourth, and fifth equilateral triangles contain 10, 15, and 21 planes.

Perhaps the five small triangles required are the five smallest triangles possible.

$$3 + 6 + 10 + 15 + 21 = 55$$

But 55 itself is a triangular number.

$$1 + 2 + 3 + 4 + 5 + 6 + 7 + 8 + 9 + 10 = 55$$

This is the answer, since the next possible answer is well over fifty.

Game 25

Game 26

There are many solutions, all finishing where five walks meet.

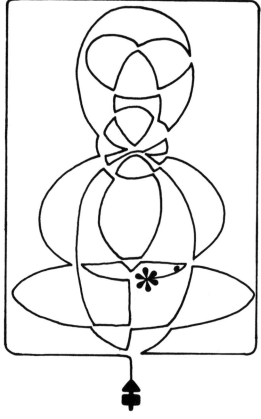

Game 27

The two loops are independent.

Game 28

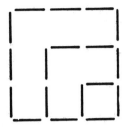

Game 29

Game 30

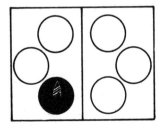

Game 31

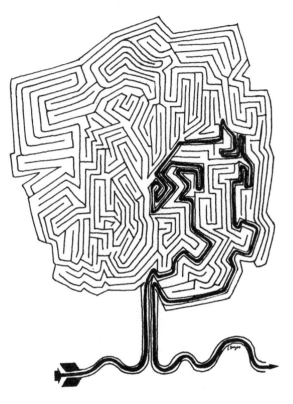

Game 32

Classify the triangles by the number of elementary triangles they contain.

1	triangle	36
4	triangles	21
9	triangles	11
16	triangles	6
25	triangles	3
36	triangles	1
		78

Game 33

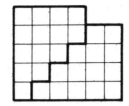

Game 34

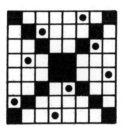

Game 35

The work can be done in six operations—two horizontal cuts and four vertical ones—leaving the pieces of wood where they were at the beginning.

Rearranging the sawed pieces won't lead to less than six operations, since the central cube needs six cuts for its six faces.

Game 36

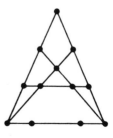

Game 37

Each couple of loops is independent, but all three are linked, like Olympic rings. If one loop is cut, the other two become free.

Game 38

The four lots are fenced off with eleven matches.

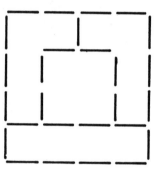

Game 39

This tour is impossible, for:

- at every crossroad but one, an even number of roads meet
- at one crossroad, seven roads meet. You can use the seventh road to enter or leave, but not both.

Game 40

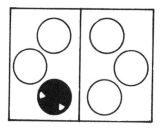

Game 41

Game 42

Classify the rectangles by the number of elementary rectangles they contain.

1 rectangle	9
2 rectangles	12
3 rectangles	6
4 rectangles	4
6 rectangles	4
9 rectangles	1
	36

Game 43

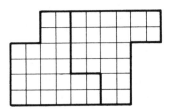

Game 44

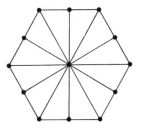

Game 45

It is important to note that the man is not pedaling while sitting on the bicycle, but pulls the pedal backward *with respect to the ground.*

If he was sitting on the bicycle the pedal and the wheels would turn clockwise and the bike would go forward.

But here, because of the gears, when the pedal turns by a small angle, the back wheel turns by a greater angle in the same direction. (The displacement of a point on the wheel is greater than the displacement of the pedal, since the radius of the wheel is greater then the radius of the pedal.) Starting from a vertical pedal position, if the man is on the ground the bike will go backward when the pedal goes backward.

Game 46

Only five moves are necessary.

```
T T T T T H H H H H
T . . T T H H H H H T T
T H H T T H H . . H T T
T H H T . . H T H H T T
T H H T H T H T H . . T
H T H T H T H T H T
```

Game 47

There are four loops. Two are free; two are interlocked.

Game 48

Game 49

At four points, an odd number of walks meet. Two of them can be used to start and finish the tour. The other two must be made into meeting points of an even number of walks. This is done by linking them with a new walk.

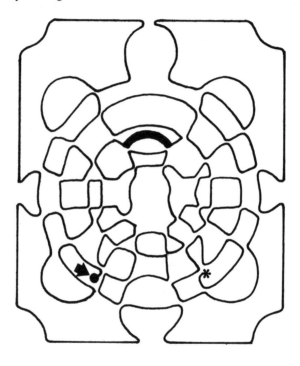

Game 50

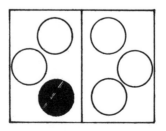

Game 51

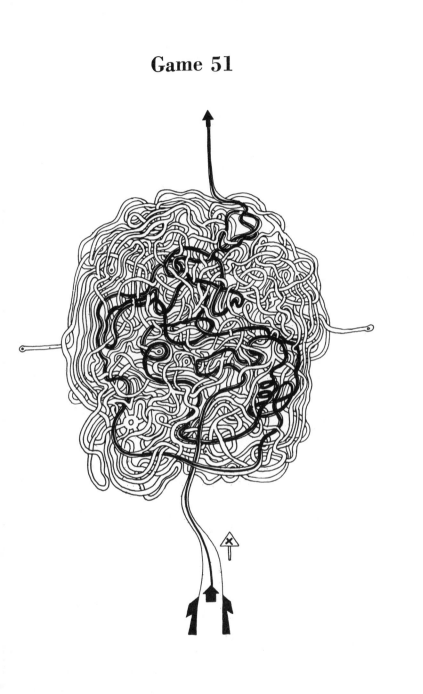

Game 52

Classify the triangles by the number of elementary areas they contain.

1 area	10		5 areas	5
2 areas	11		6 areas	2
3 areas	8		7 areas	2
4 areas	4		8 areas	1
				43

Game 53

Classify the rectangles by the number of elementary rectangles they contain.

1 rectangle	36	12 rectangles	34
2 rectangles	60	15 rectangles	16
3 rectangles	48	16 rectangles	9
4 rectangles	61	18 rectangles	8
5 rectangles	24	20 rectangles	12
6 rectangles	52	24 rectangles	6
8 rectangles	30	25 rectangles	4
9 rectangles	16	30 rectangles	4
10 rectangles	20	36 rectangles	1
			441

Within each class the count is systematic. For example, "2 rectangles." There are five ways of choosing two consecutive rectangles along a row of six; six rows mean thirty ways of choosing two horizontally consecutive rectangles. Another thirty ways for vertically consecutive rectangles (calculated the same way) makes sixty ways in all.

Game 54

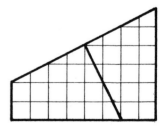

Game 55

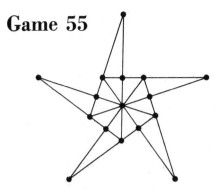

Game 56

Unexpectedly, we shall resolve this problem mathematically.

Let us represent each glass by the number $+1$ if it is right side up and -1 if it is upside down. Thus, we change one sign when we reverse one glass.

We start with

$$+1 \quad +1 \quad +1 \quad +1 \quad +1 \quad +1 \quad +1 \quad +1 \quad +1$$

The product of these nine numbers is $+1$.

The second position is represented by

$$+1 \quad +1 \quad +1 \quad -1 \quad -1 \quad -1 \quad -1 \quad -1 \quad -1$$

the product of which is still $+1$.

In fact, since each move reverses six glasses and changes six signs, an even number, the product will always be $+1$.

However, our goal is

$$-1 \quad -1 \quad -1 \quad -1 \quad -1 \quad -1 \quad -1 \quad -1 \quad -1$$

the product of which is -1. Therefore, you *cannot* get all the glasses upside down with the moves allowed.

Game 57

There are six loops. Three are interlocked with one another, and two are interlocked with each other. The last loop is free.

Game 58

You need twelve knights.

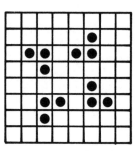

Game 59

Nine matches are removed.

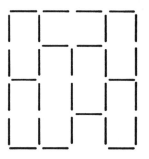

Game 60

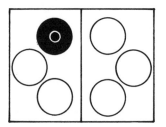

Game 61

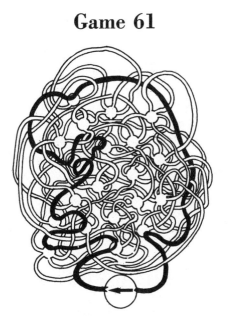

Game 62

The quadrilaterals can be convex (top), concave (bottom right), or crossed (bottom left). Count by shape.

convex	5
concave	5
crossed	5
	15

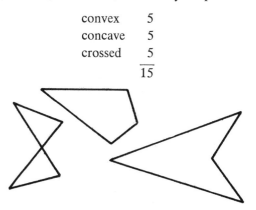

Game 63

Classify the quadrilaterals (convex or crossed) by the number of elementary triangles they contain.

Convex quadrilaterals		*Crossed quadrilaterals*	
2 triangles	4	2 triangles	12
3 triangles	16	3 triangles	8
4 triangles	16	4 triangles	6
5 triangles	8	8 triangles	2
6 triangles	20		28
7 triangles	8		
8 triangles	5	In all, there are	
16 triangles	1	106 quadrilaterals.	
	78		

Game 64

There are three loops. One is free, and the other two are knotted and interlocked.

Game 65

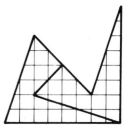

Game 66

No energy is lost to friction. The monkey and the weight are submitted to the same forces and balance each other perfectly. They climb together toward the pulley at the same speed.

Each second the monkey goes forty centimeters farther *on the rope*. But in relation to the ground, the monkey and the weight each rise twenty centimeters.

The monkey goes up at twenty centimeters per second.

Game 67

Four straight lines do the job.

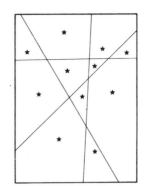

Game 68

You need twelve more knights.

Game 69

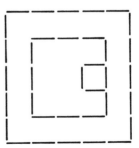

Game 70

Game 71

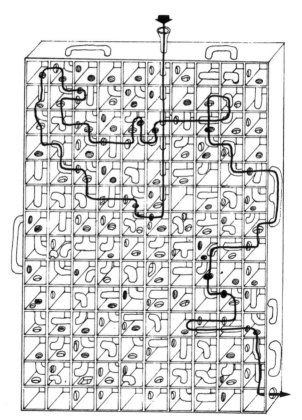

Game 72

Classify the triangles by the number of elementary areas they contain.

1	area	4
2	areas	6
3	areas	3
4	areas	3
7	areas	1
		17

Game 73

The solution comes by rotating the right angle until one end touches the midpoint of the arc.

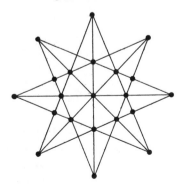

Game 74

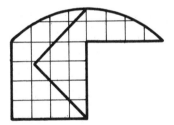

Game 75

Six straight lines do the job.

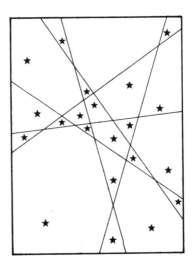

Game 76

There are four loops, forming two pairs. Each pair is interlocked and contains one loop knotted twice.

Game 77

You need eight bishops.

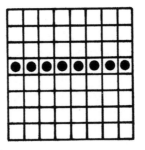

Game 78

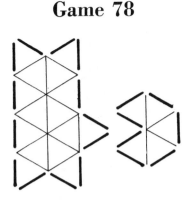

Game 79

Starting on a white square, the knight will make its first move to a black square, its second move to a white square, and so on. Every odd move will be to a black square and every even move to a white square.

To tour the remaining twenty-four squares and come back to the first, the knight must make twenty-five moves. The twenty-fifth move is odd and lands on a black square. But the corner square is white, so the tour is impossible.

Game 80

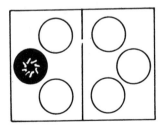

Game 81

Game 82

Regular convex hexagons		Irregular convex hexagons	
6 triangles	7	10 triangles	12
24 triangles	1	13 triangles	6
	$\overline{8}$	14 triangles	6
		16 triangles	6
		19 triangles	6
			$\overline{36}$

Concave (noncrossed) hexagons

Each of the 44 convex hexagons can have one reentrant angle in six different ways, yielding $6 \times 44 = 264$ concave hexagons.

Each convex hexagon of 10, 14, and 19 triangles can have an additional reentrant angle in one way only, yielding 24 concave hexagons.

Each convex hexagon of 16 triangles can have an additional reentrant angle in two different ways, yielding the 12 concave hexagons.

In all, there are 344 hexagons.

Game 83

There are three loops, all interlocked with one another. One is knotted once, another twice, the third three times.

Game 84

In a given time, Urban runs seventy-five meters while Vincent runs a hundred.

In another amount of time, Timothy runs eighty meters while Urban runs a hundred. Cut this proportionally so Urban runs only seventy-five meters; Timothy then runs sixty.

Thus when Vincent runs a hundred meters Timothy runs sixty. Vincent must give Timothy a forty-meter head start.

Game 85

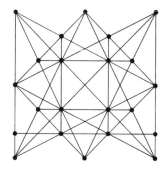

Game 86

■	▲	●	◆	★
●	◆	★	■	▲
★	■	▲	●	◆
▲	●	◆	★	■
◆	★	■	▲	●

Game 87

You need ten bishops.

Game 88

The area of a parallelogram is the product of its base and its height. Let the base be four matches long; then the height must be 1.5:

$$1.5 \times 4 = 6$$

Slide the top four matches (which are 1.5 matches above the bottom four) sideways until the slant sides are exactly two matches long.

Game 89

There are five loops. One is knotted twice and interlocked with a second loop, which is knotted once. The three remaining loops are knotless but interlocked with one another.

Game 90

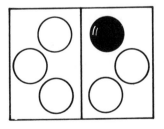

Game 91

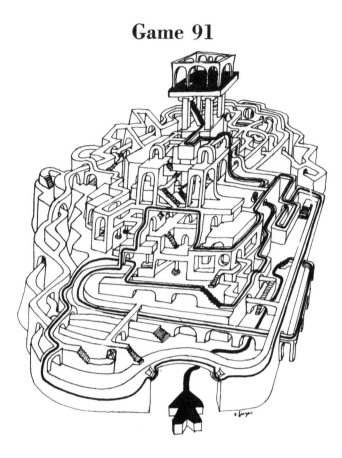

Game 92

Two loops are interlocked and knotted together, then interlocked at both ends with a third loop.

Game 93

Game 94

The forty-eight remaining squares include twenty-three white and twenty-five black squares. But a domino always covers one white and one black square. If you manage to place twenty-three dominoes there will be two squares still unfilled. The last domino cannot cover them, so no solution is possible.

Game 95

There are several solutions. Here is one.

B	B	B	B	Q	R	B	B
	Kt	R	Kt		Kt		Q
Kt	R	Kt	Q	Kt		Kt	B
Q	Kt		Kt	R	Kt		B
B		Kt		Kt		Q	R
B	Q		Kt		Kt	R	Kt
Kt		Kt	R	Kt	Q	Kt	
R	B	Q	Kt	B	B	B	Kt

Game 96

Convex quadrilaterals		
1	area	3
2	areas	6
3	areas	2
4	areas	4
5	areas	2
7	areas	2
10	areas	1
		20

Crossed quadrilaterals		
2	areas	5
3	areas	4
4	areas	5
		14

Concave quadrilaterals		
3	areas	3
4	areas	2
6	areas	2
		7

In all, there are 41 quadrilaterals.

Game 97

There are three pairs of interlocked loops. In one pair, one loop is knotted three times. In another pair, one loop is knotted twice. In the third pair, each loop is knotted once.

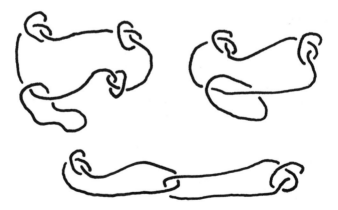

Game 98

The diagonal enters a new tile at the beginning and each time it crosses a horizontal or vertical line. But when the diagonal crosses the corner of a tile it is crossing two lines but only entering one tile. Such corners are corners of rectangles proportional to the whole floor; the diagonals of such rectangles are on the main diagonal. The number of such rectangles equals the greatest common divisor of 231 and 93, which is 3. (We include the whole floor as such a rectangle, since the diagonal, reaching its end, does not enter a new tile.)

The number of crossed tiles is:
$$231 + 93 - 3 = 321$$

Game 99

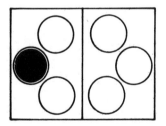

Game 100

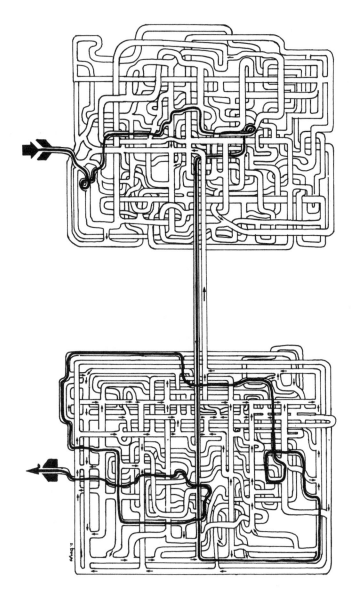